Glenn Haege's Complete Hardwood Floor Care Guide

How to Refinish & Care

for Your Wood Floor.

Edited by Kathy Stief
Illustrated by Ken Taylor
Front Cover photo by Ed Noble and Paynter Floors
Back Cover photo by G & G Floor Company

MASTER HANDYMAN PRESS, INC.

Glenn Haege's Complete Hardwood Floor Care Guide

Edited by Kathy Stief

Published by:
> Master Handyman Press, Inc.
> Post Office Box 1498
> Royal Oak, MI 48068-1498

Copyright © 1995 by Glenn Haege and Kathy Stief
> First Printing May 1995
> Printed in the Unites States of America.
Library of Congress Cataloging in Publication Data.

Haege, Glenn
> Hardwood Floor Guide; how to finish, refinish and maintain your hardwood floor.
> Bibliography: h

ISBN 1-880615-38-X

Acknowledgments

This book could not have been written without the help of a great many good friends in the floor refinishing field. From Walter Whitley, Executive Vice President of the Oak Flooring Institute, to professional hardwood floor refinishers like Bob Grabowski and Paul Maskill at G & G Floor Company, and Jim Moody at Paynter Floors, Inc., to distributors like Dick Walters at Erickson's Flooring & Supply Co., Inc., to retailers like Bill Damman at Damman Hardware, Jack Smith at Dillman & Upton, Randy Hobbs at Crandall-Worthington Co. and Doug Winfield and Gordy Winfield at Pontiac Paint.

All these professionals have been more than generous with their time, because they all want one thing: you to be successful with your hardwood floor refinishing project.

As always, the staff of WXYT Radio, Talk Radio in Detroit, have given me every assistance. The book would have been impossible without my publishing team, Kelly Boike, Barbara Anderson, Editor, Kathleen Stief, Artist, Ken Taylor of Art Staff, Page Layout Editor, Gordon Sommer, of Producers Graphics, and Doug Beauvais and Debbie Wichter at Data Reproductions.

Thanks to my family, Barbara, wife and principal proof reader, and Eric and Heather. And, as always, thanks to you, my loyal listeners and readers.

Contents

Contents

Contents

Introduction

Broadloom carpeting became popular in the 1950's. Soon every homemaker worth her salt, needed wall to wall carpeting. Today, a whole new generation of homeowners are moving into second and third generation homes and discovering beautiful hardwood flooring hidden beneath the carpeting.

America has rediscovered the strength, clean lines, and character of hardwood floors and fallen in love. Now, house and condominium owners and apartment dwellers, alike, are clamoring for hardwood floors.

They are also calling for information on how to bring the hardwood finish back to its former beauty, or create a new finish for their floors. This booklet should give the answers to most of your hardwood problems. It is dedicated to helping you bring back the warm glow of hardwood to your lives.

Many times, when the carpeting is removed, the hardwood floors just have to be brightened. Sometimes they have to be completely refinished. Other times, new floors have to be put down and heavy is the heart of the person who finds only plywood under the carpeting.

The book is divided into seven sections: **New Floors, Brightening, Finishing & Refinishing, Pre Finishing Preparation, Refinishing Procedures, Floor Care & Maintenance,** and **Problems.** The **New Floors** section is to give you the information you need to be a knowledgeable hardwood shopper. The other sections tell you how to do the job.

WARNING - DISCLAIMER

This book is designed to provide general hardwood floor care and maintenance information for the home handyman and woman. It is sold with the understanding that the publisher and author are not engaged in rendering legal, or other professional services. If expert assistance is required, the services of competent professionals should be sought.

Every effort has been made to make this text as complete and accurate as possible, and to assure proper credit is given to various contributors and manufacturers, etc. However, there may be mistakes, both typographical and in content. Therefore, this text should be used only as a general guide and not as the ultimate source of information. Furthermore, this book contains information current only up to the date of printing.

The purpose of this book is to educate and entertain. The author and Master Handyman Press, Inc. shall have neither liability nor responsibility to any person or entity with respect to any loss or damage caused directly or indirectly by the information contained in this book.

WARNING - DISCLAIMER

New Floors

1. What type of wood should I choose?

Common choices are oak, birch, hard maple and pecan. Some beautiful foreign hardwoods, like Jarrah and Jatoba are also available. Some parts of the country use soft woods like pine. Imported Russian Pine is also gaining in popularity in some parts of the country. The choice is up to you. To my mind, American hardwoods can not be beaten for beauty, longevity, and value.

Just as important as the species of wood you select, is the quality of wood from which your floor is made. The Oak Flooring Institute (OFI), an affiliate of the National Oak Flooring Manufacturers Association (NOFMA), has developed official flooring grading rules for NOFMA Certified floors[1].

Their guidelines state that Unfinished Oak Flooring be made of the top four grades, Clear, Select, #1 and #2. Sometimes the top two grades are combined into a batch called "Select and Better".

Unfinished Hard Maple, Beech, Birch and Pecan are divided into three grades: First, Second and Third. The top two grades are commonly combined into a batch called "Second and Better".

The top grade of Prefinished Oak Flooring that is readily available is called Standard Grade. Another grade of Prefinished Oak is called Tavern Grade, because of its rustic

---

[1]Official Flooring Grading Rules, OFGR/Revised, Effective November 86, Oak Flooring Institute, affiliate of NOFMA, National Oak Flooring Manufacturers Association, P.O. Box 3009, Memphis, TN 38103; p 8-9

appearance. It is sometimes combined with the Standard and Better. These bundles are called "Tavern and Better". The same is true for Prefinished Beech and Pecan Flooring.

In almost all cases, if you want the very top grade of these hard woods, Unfinished Oak Clear Plain or Clear Quartered, Unfinished First Grade White Hard Maple, First Grade Red Beech and Birch, First Grade Red or First Grade White Pecan, or Prime Grade Prefinished Oak Flooring, they have to be Special Ordered. Interestingly enough, Tavern and Better Prefinished Oak and Prefinished Beech and Pecan flooring must also be special ordered.

2. Should I choose pre-finished or unfinished hardwood?

Both have advantages, both are beautiful. The choice is up to you.

You can have a much wider selection of wood and have more control over the final look of the floor by having it finished on site. The wood you select can be stained to your exact specifications.

On the other hand, factory finished hardwood is manufactured under rigid quality control conditions by firms such as Boen, Bradley, Bruce, Chickasaw, Hartco, Robbins, and others. Bruce, Hartco, and a few smaller firms also use sophisticated high pressure techniques to create impregnated laminated hardwood flooring with exceptional beauty and wear characteristics.

Although prefinished hardwoods are initially more expensive than unfinished hardwood, much of the difference in cost is eliminated because the wood does not have to be finished on site.

There are many quality national and regional brand names. I do not know them all. Just because a brand is not listed here does not mean that its product is not of the highest quality. Some manufacturers make their flooring available in both factory finished and unfinished forms.

3. What are the differences between factory finished and job finished hardwood floors?

On the plus side, factory finished hardwood flooring has a much stronger and more durable finish. Factory finished hardwood floorings are often laminates, with the top hardwood layer attached to two to four layers of wood. These engineered woods offer a more stable foundation and add a range of use, that would be impossible for natural hardwood. Factory finished flooring is also quicker to install.

On the minus side, factory finished hardwood flooring has a more restricted selection of woods, is slightly more expensive, and, if an accident happens, is somewhat more difficult to repair.

Job or site finished hardwood flooring is less expensive and gives you a broader selection of woods and finishes. The beautiful imported hardwoods like Jarrah, Jatoba, and the different rosewoods, are only available as unfinished lumber. If the wood is finished on site, you can have any finish you desire.

On the minus side, job finished hardwood flooring, shows wear faster and is more easily injured. However, when the finish is injured, repair is easier.

4. Can hardwood floors be laid on any surface?

A hardwood floor can be ruined if it is installed on the wrong surface. Hardwood floors must be laid on clean, solid, flat, dry surfaces. Anything else is an invitation to disaster. Water and hardwood do not mix. Most hardwood floors will be ruined if laid in basements or other damp areas.

If you want to lay a floor on a concrete slab, be sure to test the concrete for moisture before you lay the floor. New concrete has to cure for a minimum of sixty days up to two years before it is dry enough to lay hardwood. Also remember that a waterproof barrier, such as polyethylene film, asphalt paper felt, or vinyl moisture barrier, may have to be placed between the slab and the hardwood.

Prefinished flooring has a definite advantage in problem areas. Many are specifically engineered to stay strong and beautiful under very harsh conditions.

5. Can I install a hardwood floor in the basement?

The rule always was that wood floors were not to be laid below grade because of the moisture problem. Now a number of manufacturers such as Boen, Bruce, Hartco and Robbins have heard your pleas and developed revolutionary prefinished hardwood surfaces, methods of sealing concrete slabs to block moisture and other installation techniques that permit some of their brands to be laid below grade.

6. Can I install a hardwood floor by myself?

Yes you can. My hardwood floors were all installed by an expert hardwood flooring contractor when the house was being built. He did a spectacular job. But that doesn't mean that I couldn't have installed them myself, or that you couldn't.

Some parquet hardwood flooring is no more difficult to install than vinyl tiles. There is a lot of work involved and you have to be very meticulous. But if you've got the time and energy, and you're willing to follow directions to the letter, you can do it yourself and gain full bragging rights.

If you're going to do the job yourself, you don't have to go it alone. You can get very excellent instructions from NOFMA Technical Service, the Oak Flooring Institute of the National Oak Flooring Manufacturers Association. They produce ***Installing Hardwood Flooring, Finishing Hardwood Flooring,*** and the ***Wood Floor Care Guide.*** I recommend all of them.

Here's the address and phone number:

National Oak Flooring Manufacturers Association
P.O. Box 3009, Memphis, TN 38173-0009
Phone: 901-526-5016.

Many manufacturers produce their own easy to follow installation manuals for their various lines. Some even have videos. On the local level, more and more wood flooring suppliers have free flooring installation clinics that are open to the Do-It-Yourselfer. You owe it to yourself to go to one or two of these clinics if you are even thinking of installing a hardwood floor.

7. Can I install a hardwood floor in my kitchen?

I want to replace a vinyl floor. We have children, dogs, and an outside pool so it will receive a great deal of abuse. Can I still have a hardwood floor or must I stay with vinyl?

You can have a wooden kitchen floor. I would suggest that you consider installing either a factory finished floating floor or an impregnated hardwood factory finished floor.

A floating floor is a method of laying a floor in which they put a very thin ethylene foam pad on the floor, then specially constructed hardwood panels are put down. During manufacture, each prefinished hardwood flooring strip is laminated to a same size piece of plywood. In other words the prefinished hardwood is glued to another wood strip, wood to wood.

This is very firm, rugged construction. On the job site, the hardwood/plywood combination units are glued tongue to grove, then laid down on a special foam base. No mastic is applied to the foam. The rigid hardwood flooring becomes a very tough single unit that "floats" above the subfloor on a thin cushion of foam.

The floating floor is often used where the contractor cannot make the subfloor perfectly flat. It also adds insulation, is more resilient, and acts as a sound barrier. Installation is very simple because the hardwood strips just have to be cut to size and glued together. No nails are needed.

Another choice for a high traffic, hard use area is impregnated hardwood flooring like Hartco's Pattern-Plus®. This is also hardwood that has been finished at the factory. It comes in boxes and all you have to do is glue it down. A big advantage of this type of flooring is speed. Although more expensive than traditional flooring, pre-finished flooring takes only one third the time to install than unfinished hardwood flooring.

Impregnated hardwood can take a lot of heavy use because the stain and protective sealer has been impregnated deep into the cell structure of the wood. This is done in the factory by creating an artificial, highly pressurized, atmosphere and forcing the stain and finishes throughout the fibers of the wood. This makes the entire hardwood layer a 120 mill deep wear surface.

For the record, unfinished floor adherents would tell you that it is better to install a wood floor, stain and coat it with a heavy duty commercial finish like Street Shoe, and screen and refinish the floor every three to five years.

8. When should a hardwood floor be finished in a new house?

Ideally floor finishing should be done within one to three weeks after installation. However, the finishing process should not be started until all other work in the house has been completed. That means all the cabinets have been installed; the painting has been finished and just before the final coat of finish has been applied to the base molding.

Cleaning prior to finishing should be sweeping and vacuuming only. No water should be used on the raw wood floor. If hardwood flooring is desired on a concrete slab, let's say in a family room, it can not be laid until the concrete has cured.

Once the floors have been finished, the house should be kept at or near the temperature needed for an occupied house. Major temperature fluctuations can affect finish curing and stress the wood.

Brightening

The easy way to renew the finish of a hardwood floor.

Light Refinishing or Brightening is the most important tip in the book. It is so important that I am putting it in front of any other sanding or refinishing instructions.

Read this tip and then look at your floors. If the finish is old, dull and yellowed, but not ruined or worn through to the bare wood, try this before completely refinishing.

Do one room. If it works, and it usually does, you have saved days of very hard work and hundreds of dollars. If it doesn't, never fear. The rest of this book is filled with detailed instructions, and I will be with you every step of your refinishing adventure.

9. How do we bring back the luster to our hardwood floor?

When we tore up the carpeting we found that the hardwood floor was in fairly good condition. Just a little dull and old looking.

The average homeowner who pulls up the carpeting and finds a hardwood floor has a site finished floor. That is the wood was finished on the job site. The following instructions are only good for site finished floors. Floors with factory finished flooring can usually be renewed with special wood restoration mixtures distributed by the flooring manufacturer.

Most floors do not need to be refinished more than once every 20 or 25 years. Usually, the top coat of urethane has just gotten old and yellowed. While adding new coats of the old oil based polyurethanes was often not successful, the new VOC finishes make it possible to add a youthful new luster and years of extra life to your floor's finish. Here's how to do it.

Materials Needed: TSP, fine sanding screen for 16" buffer or fine grit sandpaper for Flecto SQUAR BUFF orbiting buffer, Verathane Elite Diamond, Faubulon Professional Crystal II and additive, or other VOC urethane finish.

Equipment Needed: 16" buffer or Flecto SQUAR BUFF orbiting buffer, sanding respirator, wet/dry type vacuum with a bag and attachments, broom, lambs wool pad applicator and extension pole.

This is an easy, fast job. You can be done in a day. Sanding creates a lot of sanding dust. The job goes twice as fast if you can make it a two person job. One sands, the other vacuums continuously.

Everyone in the room should wear a respirator when sanding is going on. There will be a great deal of fine dust in the air, so turn off pilot lights and do not allow smoking in the house during the sanding operation.

1. Remove everything from the room. Take up the quarter round moldings. **Turn off the furnace when sanding and finishing the floor. Tape shut the forced air vents and returns.**

2. Clean the floor thoroughly. There should not be a hint of grease or wax when you are done. If there is any wax on the floor, it must be stripped.

Wash the floor with a mixture of 4 oz. dry measure of TSP per gallon of warm water. Rinse several times. Let dry thoroughly.

3. When everything is clean and dry, put on a respirator and lightly sand or screen the floor. Just cut the top surface. If you can find a Verathane Flecto dealer, renting their SQUAR BUFF vibrating sander would be perfect for this job. Many home centers carry the line. In addition to renting the sander, you would need to get 80 grit paper.

You can also use a large 16" floor buffer. Attach a 3M Brand Fine Grade, round sanding screen and buff lightly. Just sand until a light powder appears over the entire surface.

Sand all edges, corners, closet floors, everywhere you could not get to with the big equipment with a hand-held orbiting sander, finishing sander or drywall sanding screen.

4. Vacuum thoroughly, floors, walls, all ledges, everything. It is important to use a wet/dry vacuum with a bag because the fine sanding dust can ruin the motor.

Do not allow any dust in the room. Vacuum everything: floors, walls, window sills, doors, everything.

5. Wipe down the floor with a slightly water dampened rag attached to a broom. Repeat if necessary. Do not let even a hint of dust remain in the room.

6. Pad on a couple of coats of VOC Flecto Verathane, Fabulon Professional Crystal II, or other VOC crystal clear finish. Pad the coats on a couple of hours apart.

This procedure should make your floor look brand new. If you use Fabulon be sure to get the additive or it will not adhere to the old surface.

One word of caution: Don't make this project any bigger than it is. Do not sand down to the bare wood. The only reason for screening the surface is to create a profile that will enable your new finish coats to bond to the existing finish. Depending upon wear, you should only need to drum sand a floor once every 20 to 50 years.

You're done. Brag about it!

Finishing & Refinishing

10. What type of finish should I choose?

Basically, there are two different types of finishes you can put on a hardwood floor: penetrating sealers and surface finishes. Penetrating sealers are softer and more mellow. Surface finishes have a wide variety of sheen rates.

There is a third class of finish called irradiated polymers. These are very complicated and used in commercial applications and will not be covered here.

Penetrating Sealers

With a penetrating sealer, the product soaks into the wood pores and hardens to form a protective seal against dirt and stains. Since the seal has soaked into the wood it wears only when the wood wears and traffic patterns are not obvious from surface wear.

This is the type of finish that is recommended for residential use by the Oak Flooring Institute (OFI). They go even further and state that nonprofessionals should use the normal or slow drying penotrating sealer. They fear that the nonprofessional will have to work so rapidly with the fast drying sealer that he/she may ruin the looks of the job.

Penetrating sealers are very forgiving. Wear patterns often can be eliminated with the use of professional cleaning and reconditioning compounds made by Hartco, Minwax Duraseal Division, Bruce and many of the major coatings suppliers. If the pattern still exists it can usually be eliminated by applying another coat of sealer. You have to check with the individual manufacturer's instructions to see if recoating is recommended.

Surface Finishes

Surface finishes, as their name implies, form a protective, often satin or high gloss, layer of protection on top of the wood. The five different types of surface finishes used on hardwood floors are Polyurethane, Varnish, Shellac, Lacquer, and VOC finishes. Of these, Polyurethane and the VOC Finishes, provide the strongest finish and are the least temperamental.

Oil Based Polyurethanes:

Oil Based Polyurethane finishes were the standard of the industry for many years. They provide a very strong but brittle, golden hued finish. They do not film build well and release distinctive "paint fumes" into the air during the first 3 or 4 hours of drying time. Although dry to the touch in two hours, you usually have to wait overnight to recoat. In a relatively short time, usually only one or two years, the finish takes on a golden hue.

VOC Water Based Finishes:

The newest surface finishes are water born VOC urethane and/or acrylic combinations. VOC stands for Volatile Organic Compounds. VOC Compliant Finishes, often called VOC finishes, conform to strict new air quality regulations and release little or no odor or potentially harmful solvent compounds into the air during drying.

VOC finishes are very thin and at least four coats are required to create the necessary film build. Their need for multi-coats is offset by the fact that drying time is very short and new coats can be applied every one and a half to two hours. A VOC finish is crystal clear, and will not yellow in one or two years like oil base polyurethane.

The number of high quality new VOC finishes is growing monthly. Soon every manufacturer of floor finishes will have a VOC product. Here is a partial list of some, but by all means not all, of the new finishes: Carver Tripp Safe & Simple, Coronado Aqua Plastic, Diamondlac Rawhide, Fabulon Professional Crystal II Clear Acrylic, Minwax Polycrylic, Pacific Plus, Pacific Strong, Satin GymSeal by McClosky, Sikkens TFF, Street Shoe by Basic Coatings, and Varathane Elite Diamond.

There are two reasons for this growing popularity. First, governmental air quality standards are increasingly weighted against products that release volatile organic compounds, like those found in oil and alcohol based products, into the air while they are drying. Second, the public is learning that the new VOC finishes are great DIY products. They are faster drying, easier to work with, and more pleasurable to use.

Most of the new VOC finishes are slightly softer than the oil based polyurethanes. If you have large pets and young children tracking in water and grinding in dirt, or a great deal of floor traffic, you may decide to upgrade to a super strong VOC like Street Shoe by Basic Coatings or Gymseal by McClosky.

The better VOC finishes contain less acrylic and more urethane. This means that they do not flow as easily, but are much "tougher". These high quality, professional finishes also need to have a catalyst added before application. The catalyst contains Aziradine and is purchased separately. It makes it possible for the very smooth coats of finish to adhere to one another. It also adds chemical and abrasion resistance. Extra care must be used when adding Aziradine. Don't spill. People with sensitive skin may have a reaction. Make sure stroage, application, and trash areas are kid and pet proof.

Take the time to write down your requirements. Describe the abuse your floors will have to take in detail. Take your requirement list with you when you go finish shopping. Make sure your retailer proves to you that the finish you are buying meets your requirements.

This is especially important because all of these top of the line finishes are **very expensive**. The VOC finishes are commonly 30 to 200% more expensive than the oil based polyurethane and usually require more coats of finish. However, they cover more square feet per gallon, do not release hydrocarbons into the atmosphere, are more forgiving, have water clean up, and all and all, are easier and safer for the DIY-er to use.

11. Do I need to put wood hardener on a hardwood floor.

No. A hardener is only necessary with a soft wood like pine. It penetrates the surface and hardens, making the softwood accept stain more like a hardwood. You already have the best that mother nature can provide.

12. Can I save a little money on my new house and finish the floors myself?

I'm having a new house built and will be installing hardwood flooring throughout.

Yes you can! As long as your mortgage lender and builder agree, you can save some money by painting the walls and finishing the floors yourself. I didn't. When I built my new house I was delighted to have top flight professionals do the painting and floor finishing. I considered it money very well spent.

Deciding whether you can/should do this sort of thing on a brand new house is a very complicated, personal, time/

money/need/want calculation. If you need to save money, or if you have the time and want to do a good bit of the work yourself, there are a lot of things you can do. You can string the wire, install the insulation, paint and wallpaper the walls, and finish the flooring.

On the other hand, taking possession of a new home is a frenzied time in any homeowner's life. There are many things you have to approve, or shop for on the new construction. You have to get ready to move. There are always last minute hassles on the financing. I guarantee that you will feel like a dog chasing its tail in the last few weeks. Increasing the chaos by finishing the floors may be too much to handle.

Professional floor finishers can apply special products, like the Glitza lines of finishes, that you can not even buy, yet alone apply. All things being equal, there are a lot of good reasons to let the professionals do it. But, if the will is there, you can do it yourself and save. This book has the information you need to have a professional looking job when you are done.

13. Professional Vs DIY

We are going to take the carpeting out and have the hardwood floors refinished. What would you recommend?

Floor finishing and refinishing is a very high skill occupation. It is also a trade in which caveat emptor, let the buyer beware, is a very real admonition. The true craftsman can make magic happen. Anything less can be a horror story.

The pros in the business tell me that it takes one and a half to two years to train a craftsman. It is very hard work and a highly meticulous profession. Most people do not have the temperament to do the work properly.

If you are in the market, it is up to you to investigate the contractor. Anybody with a sander can say he or she is in the business. Ask for recommendations. Call, and go look at several of his or her jobs.

This part of the book may sound sexist. I know a good many ladies who have done a bang up job brightening or renewing their hardwood floors. However, a good professional drum sander can weigh up to 320 pounds. These heavy sanders do a much better job, faster, flatter, than the 100 - 130 pound buffers. Not many women go for the gusto with a sander that weighs over 250 pounds, so professional finishing/refinishing is a predominantly male occupation.

A good finisher/refinisher has top of the line, beautifully maintained equipment. Anything less and you are talking to the wrong guy.

14. What type of finish should I specify?

When you contract for a floor to be professionally finished or refinished, you are buying a combination of skill and artistry. You would never tell a heart surgeon how to do the job or what brand pace maker you want installed, if the need arises (and I hope it won't) you will be buying the heart surgeon's expertise and want him to do what he/she feels best.

The same is true with a floor refinisher. You have to investigate enough to make certain that he gets good results. You may want specific properties in your newly refinished floor, such as wear strength or water resistance.

List the properties that are most important to you, then let your professional choose the product that will best fill those requirements. As a professional, he has a much wider range of products available to him than are available on the retail market. The contractor should work with the product with which he gets the best results.

If you have studied the finishes and want to specify, find a finisher who likes the same finish you do. One of the finishes I recommend is called Glitza.

Glitza is the name of an old line company that makes a number of different products. The Glitza based system that professionals I respect, like G & G Floor Company, have used for years. The system consists of one or two coats of an alcohol based epoxy sealer and a urethane top coat. This finish is very hard.

A professional product like Glitza must only be applied by a professional. It utilizes activators and inhibitors, and releases a number of different chemicals into the air while drying. Wood refinishers using the Glitza system wear charcoal filter respirators, and OSHA is talking about requiring them to use self contained compressed air.

That said, the Glitza finish system is one of the truly beautiful, long wearing hardwood floor finishes on the market. The final finish coat takes 12 to 24 hours to dry and curing will go on for 30 to 60 days depending upon heat and humidity.

15. Is there anything I can do to help the contractor when the floors are being refinished?

Yes, take your entire family and go away. No people or pets should be near the surface for as long as possible. The floor will be dry to the touch within 12 to 24 hours. But leaving it alone for three or four days will give you a better, tougher, finish. Actually the more expensive the VOC finish, the lower the odor level. With some finishes, a floor could be completely refinished and there would only be a slight ammonia odor, similar to that which you would have with latex paint.

Professional finishes release a great many hydrocarbons or alcohol into the air when drying. Try to have someone open windows and let the air circulate as soon as the finish has dried to touch.

With the sanding and all the chemicals being released into the air, your air filter will have been working overtime. If you have an electronic filter system, clean it immediately after the work has been done.

If you have replaceable furnace filters, install a new filter. Have it ready before the work begins, so that you can put it in after they are done. That way you will be sure to filter any chemicals that are oxygenating (flashing off) the first 24 hours.

16. Can I refinish the floors myself?

Yes you can! The manufacturers and retailers have really been listening to the demands of today's Do-It-Yourselfer. New equipment and finishes on the market make floor refinishing far easier than it was just five or ten years ago.

That being said, remember, floor refinishing is not for everyone. You should be in good physical condition and really want to do it. There is a great deal of hard work involved. You will learn that you have muscles you didn't even know existed before. You have to be very meticulous. You cannot rush.

The actual application of the finish is relatively easy. Surface preparation is the hard part. Only strong backs and young knees need apply! A few grains of sawdust can spoil the look of a job.

Many savvy people split the duties. They hire a professional to prepare the surface. Then, the DIY'er applies the finish. This is still a tough job.

The Good Book talks about people that have ears and hear not, and eyes and see not. If you want to refinish hardwood floors successfully, keep your ears and eyes open. Learn everything you can learn before you start. Follow directions to the letter. Don't take short cuts.

Many homeowners who want hardwood floors, already have them. They are either hidden under carpeting, or their floors have been in constant use and they are making the decision as to whether a little touch up, or a full refinishing job, is in order. The following tips should help.

17. Should my home have hardwood floors?

One final note before you spend the next four or five weekends refinishing your hardwood floors. Pause and reflect (that is a very apt word) for a minute. There is a reason that the homebound housewives of the 1950's were almost willing to kill for wall to wall carpeting.

It was not that they were bored or angry that the culture of the day prohibited them from fully expressing their individuality by going into the business world and becoming corporate tycoons. They were not taking out there frustration by smothering their floors instead of their domineering husbands. These very intelligent women wanted wall to wall carpeting because the look and feel of carpeting is warmer than hardwood, and a lot less work. Especially if you have pets or kids.

Visualize a beautiful dining room table. The wood is a work of art. Unfortunately, dust collects like a magnet and you don't feel safe unless it is protected with a linen table cloth.

You are contemplating turning the floors in one or more rooms of your house, along with high traffic hallways, into gigantic dining room tables.

Hardwood floors are a lot like children. They require constant attention. Hardwood floors should be vacuumed at least every other day. Even dust mopping can scratch the surface.

I love hardwood floors. I have hardwood floors. But before you make the leap, take a minute or two to pause and decide whether hardwood fits your lifestyle.

Pre Finishing Preparation.

18. You can't scare me. How do I get started?

> **Materials Needed:** 3" Wide Painter's Tape, water base wood filler, wax stripper if needed, TSP.
>
> **Equipment Needed:** Tools for removing quarter round molding, nail set, hammer, putty knife, buckets and mops, old 100% cotton towels.

Like painting, proper preparation is the first step.

1. Carefully take off quarter round moldings. You may want to number them if there is any possibility of a problem developing when you have to put them back in place. They should be cleaned and finished or refinished separately.

2. Turn off the furnace, water heater and all other pilot lights. Tape shut forced air registers and returns.

3. Mask the floor molding with 3" Easy Mask by Daubert Chemical or 3M Blue Tape, not regular masking tape. Three inches of protection can be a real life saver if the sander or buffing machine tries to run away from you.

4. If the floor was waxed, it must be stripped completely before sanding. Use a good commercial stripper.

5. Dirt and grease are great enemies of hardwood. The heat of sanding can actually work them into the wood. You must always start with a completely clean surface or the finish may not adhere properly.

The Oak Flooring Institute does not recommend the use of water on raw wood. However, if the floors are greasy, I believe that you may need to wash them down with a solution of 2 oz. dry measure of TSP per gallon of water. Rinse several times with fresh water.

6. Never let water stand on wood. Dry the floor continuously as you work with 100% cotton towels or rags.

7. Some moisture will have soaked into the wood. Air dry the floor for at least one day before beginning the sanding process. If there is a lot of humidity in the air or you are working in a new house that is in the process of drying out, more drying time may be needed.

General Finishing.

19. How should I prepare new hardwood floors and put down the finish coat?

I have installed 2,000 square feet of hardwood floors and want to put Fabulon Professional Crystal II on it. How should I prepare the floors and put down the finish coat? I am having a great deal of trouble getting the scratches out. I want the wood to be as close to natural as possible. Do I need to put on a stain?

Materials Needed: Wood stain, Fabulon Professional Crystal II or other VOC water based acrylic wood flooring finish, 3M synthetic steel wool, light sanding screen.

Equipment Needed: Sanding screens, vacuum cleaner, Clark or other heavy duty buffing machine, lambs wool pad applicators.

The traditional oil based polyurethanes, tended to give wood a warm golden hue. The new VOCs are crystal clear so that "warm glow" is missing. You may want to warm the wood with a light stain.

Another reason for adding a light stain is that if you have an extensive area of hardwood flooring, there are so many different angles that imperfections will tend to glare at you. A light stain accentuates the grain structure and tones down defects.

You can use these same general directions for preparing and applying any VOC (Volatile Organic Compound) finish.

1. Be sure you removed the quarter round molding, protected the wall molding with painter's tape and cleaned the floor thoroughly before you start.

2. Sand the entire area using the procedure outlined in the next tips. Be sure that when you do the final buffing of the floor with a Clark or other heavy duty circular buffing machine that you use a Synthetic Steel Wool pad or screen if you are going to use a VOC finish. If you use traditional steel wool, small steel filaments will become imbedded in the wood. These imbedded filaments will rust when a water based VOC finish is applied. This will cause little brown spots that can't be washed away because they are below the finish.

3. After you sanded the floor, vacuum thoroughly. Include the entire room, window sills, doors, ledges of any kind will hold dust. They all have to be vacuumed meticulously with the vacuum cleaner attachment.

When done, wipe down the floor with a water dampened rag that you tie around a push broom.

This is one of the most vital steps in the entire process. There should not be even a speck of sawdust left when you are done.

4. Apply the stain with a lambs wool pad applicator. The applicator should be washed in clear water and air dried to remove any lint *before* use.

Go with the grain of the wood. Plan your route so that you do not wind up in a corner you can't get out of.

5. Allow the stain to dry thoroughly.

6. The stain will have raised the grain of the wood. It is therefore necessary to repeat steps 2 and 3, lightly sanding, vacuuming and wet tacking. Be especially careful to get up every speck of dust.

7. Gently pour the finish into a 9" tray and apply the first coat of the Fabulon Professional Crystal II. I recommend that, if you do not already have one, you purchase a deep well roller tray and use a tray liner.

Use a pre-washed lambs wool pad applicator not a roller to apply the finish. Anything else will give you an imperfect finish. **Work Slowly!** VOC finishes go on like water. Many DIY'ers start working so fast they miss a spot. It is a good idea to clamp a light in a corner directed at the floor. The shine will help you identify areas you have missed when applying the finish.

8. The VOC finish will have raised the wood slightly. Screen the floor again with the buffing machine. You do not want to cut through, just raise a light powder on the surface. Vacuum and wet tack the floor by wiping it down with a water dampened rag attached to a broom. Repeat if necessary to remove all the dust.

9. Put down three more coats of the Crystal Clear Acrylic. Lightly sand, vacuum, and water tack between each coat.

There are no harmful fumes with a VOC, and you can clean up with soap and water. It is very safe.

20. What tools should I use for sanding?

There are three different sanding techniques. Each uses a distinctive set of tools. The three different systems are: the professional, the buffer and the Flecto systems. The professional uses a broad spectrum of tools: the drum sander, an edger, a 16" buffer and a vacuum cleaner. As the name implies, the buffer system uses a 16" buffer, an edger or hand held orbiting sander and a vacuum cleaner. The Flecto uses a unique SQUAR BUFF floor sander and a vacuum cleaner.

Professional Sanding Technique tools

Drum Sander

The drum sander is a heavy 200 to 300 pound plus piece of equipment designed exclusively for finishing and refinishing hardwood floors. It is the ideal piece of equipment for this job.

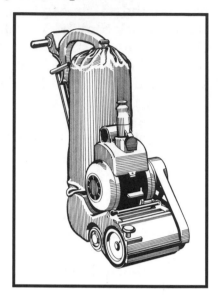

When operating a drum sander the operator starts moving the sander backward or forward, then lowers a rotating sanding drum onto the floor.

The drum grinds the surface and exhausts the sanded material into an exhaust bag. At the end of the pass, the drum is lifted from the floor before stopping or changing from a forward or backward movement. The drum sander gives the fastest, smoothest, most even sanding job.

The only problem with the drum sander is that it is a professional piece of equipment. Extreme care must be used. A drum sander must never be started or stopped with the drum lowered. If you pause for even a moment, the rotating drum will sand a furrow into the floor.

Rough, medium and fine sanding grits are used on drum sanders. The rough grade is used for the first sanding of the floor. This is called the first cut. On a new floor the first cut is used to even out any imperfections. On a floor that is being refinished, the first cut is used to remove the old finish. A special "open face" rough sanding grit is used for this so that the old finish does not clog the paper.

Medium and fine sanding grits are used on the second and third cuts. The primary purpose of this second and third sanding is to smooth out the groves left by the rough sanding grit.

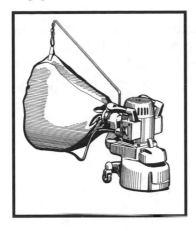

Edger

The edger is a heavy duty sander that is used to sand the edges and other small areas where it is impractical to use a drum sander.

When operating an edger, the operator uses the same sequence of sandpapers, rough, medium and fine, used with the drum sander.

16" Buffer

The 16" buffer is used to smooth out any swirl marks left by the drum sander and edger. Fine sanding screens and non metallic steel wool pads are attached to the buffer for this procedure.

Vacuum Cleaner

A heavy duty wet/dry bag vacuum is used to pick up sanding dust. Be sure to use the bag vacuum, because the fine dust can ruin the vacuum motor.

Buffer Sanding Technique tools

16" Buffer

When using this technique the 16" buffer is used to do the entire stripping job. The buffer is lighter weight, takes longer and is harder to control than a drum sander. It is better to use a buffer than a drum sander: when the wood has already been refinished several times and the boards are only 3/8" to 1/2" thick or when you only want to brighten not completely refinish the floor and only a light screening is necessary.

If you are refinishing the floor, the old finish is removed with an open face rough sandpaper disk. Then the floor is smoothed out with repeated sandings using medium and fine sandpaper disks.

Edger or Orbiting Palm Sander

An edger or hand held orbiting sander is used to get to areas that are inconvenient for the buffer: corners, edges, closets, etc.

Bag type, Wet/Dry Vacuum cleaner.

A heavy duty wet/dry bag vacuum is used to pick up sanding dust. Be sure to use the bag vacuum, because the fine dust can ruin the vacuum motor.

Flecto Sanding Technique tools

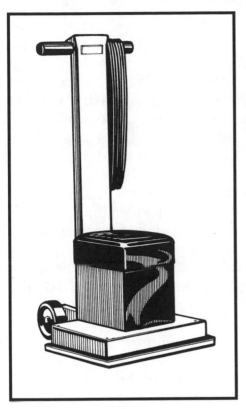

SQUAR BUFF
Floor Sander

The SQUAR BUFF floor sander is a 130 pound orbiting sander. It is almost fool proof to use. The SQUAR BUFF has a two hand operating grip, rear wheels and a large, rectangular sanding surface that uses 3M Scotch-Brite sanding pad and four grits (very coarse, coarse, medium and fine) of sanding paper.

Random Orbiting Sander

The random orbiting sander is used along edges, in corners, and in closets, wherever it is difficult to use the SQUAR BUFF sander.

21. What is the professional technique for sanding a wood floor?

Remember that I said it usually takes one and one half years to teach a craftsman how to finish and refinish hardwood floors. The drum sander can literally destroy your hardwood floor in a few minutes times. Using a drum sander the wrong way can take years of life from a hardwood floor. Extreme caution is called for. Play by all the rules, then double and triple check to make sure you are doing it right.

As I said earlier in this booklet, the Oak Flooring Institute (OFI) has published a Hardwood Flooring Finishing/ Refinishing Manual. These are the real professionals in the business and the only reason for their existence is to make certain that you have a hardwood floor that you can be proud of. I recommend their manual very highly. You will find their address and telephone number on page 13 of this book.

Before you start resanding a previously finished floor, go over to the edge of the floor where a floor heater or register has been removed and measure the thickness of the hardwood.

Most hardwood flooring is 3/4" thick and can be sanded several times. Some floors are only 1/2" to 3/8" thick and extreme caution must be used. Do not use a drum sander on these thin floors. Use a 16" buffing machine and rough, medium and fine sanding screens. Remove as little of the wood as possible.

Recommended Equipment

First, you need the proper equipment to do the job. Unless the wood is too thin, heavy duty drum sanders are the best. They are not only heavy duty, there are also very heavy, weighing from about 200 to over 300 pounds. Be sure that the sander you rent is in good condition.

You can use a 16 inch disc sander, or buffer, instead of a drum sander. All things being equal, I would rent the drum sander. Drum sanders move in a straight, level line. My advice is to use the drum sander to do the majority of the finishing or refinishing, then take out light swirl marks with the 16 inch sander/buffer and a light sanding screen.

There are three times when you should use a 16' buffing machine: 1, when that is all that is available; 2, when the wood is 1/2' or less in thickness; and 3, when there is only a very light finish on the floor.

In the last case you may be able to take off the entire old finish with only one pass of the disk sander and extra fine sanding screen.

If you cannot get a drum sander, and only have access to a 16 inch sander/buffor, or the wood is 1/2" or less in thickness, you can still refinish the floor. However you will have to use a great deal more care to get an even job. I will give some 16 inch sander/buffer use hints at the end of this section.

Make certain that the rental equipment company can provide everything you need. This includes the drum sander itself; coarse; medium and fine sandpaper to fit the machine;

a power edger and coarse, medium and fine grit sandpaper for the edger; and a 16" buffer, complete with sanding screen and Non Metallic Steel Wool pad. You will also need sanding blocks, a scraper and a respirator.

Some rental equipment companies only stock coarse grit paper. If that is the case, you may be wise to go to another rental company. If they do not stock the sandpaper necessary to do a good job, how can you expect them to give you the support you may need to complete the project? Don't even start to refinish a floor without having a supply of coarse, medium and fine grit sandpaper for your sander.

When refinishing a floor, OFI recommends that you buy open face rough sandpaper to remove the finish because the heat generated by sanding will cause the old finish to gum up and clog traditional sandpaper.

Some people make do with a light weight random orbit finishing sander, rather than renting an edger. The random orbit sander will do the job but it is a rather light weight machine. If you are going to do several rooms you will put a lot of strain on both yourself and the equipment. The edger is heavier and heavier duty. It is a little more difficult to use, but will do the job more rapidly.

One Word of Caution:

Always wear a respirator when sanding or vacumming sanding dust. Your lungs are precious. At a minimum, wear a 3M Dust/Mist/Sanding Respirator No. 8560/8710 or equivalent.

---

22. **Professional Drum Sanding Technique for new wood floor, or an old wood floor where refinishing is required.**

> **Materials Needed:** Open faced rough grit, and regular, medium and fine grit sanding belts or disks for an edger, 16" fine sanding screen and Non Metallic steel wool pad, wood filler.
>
> **Equipment Needed:** Respirator, drum sander, 16" buffer, edger, wet/dry, bag mounted, vacuum with attachments, sanding block, scraper, putty knife.

Never sand a floor with loose wood. All wood must be firmly attached to the subfloor before sanding can be started. If you have had to nail some of the hardwood strips to the floor, make sure that you countersink the nails before sanding.

When you use a drum sander you will notice that the sanding drum can be raised from the floor. Always make sure that the drum has been raised before you start or stop the sander. If you start or stop the drum sander with the sandpaper in contact with the floor, you will cut a furrow into the floor that can not be removed.

Empty the dust bag on the drum sander whenever it is 1/3 full. It is a good idea to empty it at the end of every cut.

Some firms recommend that all older floors be sanded on the diagonal. The Oak Flooring Institute, most of the professional floor refinishers I know, and I agree that except in very particular cases, sanding should be with the grain.

Start at the far right wall 2/3 of the way across the width of the room. Start the sander, begin to walk, then gently ease the sanding drum onto the floor. Slowly ease the sander along the wall to the far wall. You can always get a big charge out of electricity. Be very careful not to run over the electric cord.

If you are sanding a new floor, you just want to even out an imperfections and unevenness that may be in the new floor. If you are sanding a floor that has already been finished, you want to remove just enough to take off the old finish. The art form is to remove the finish, but barely touch the wood. Keep watch constantly to make sure that you are not sanding away too much wood.

Just before you have reached the far wall, lift the sander from the floor. Start pulling the sander straight back in a straight line to its original position. Gently lower the sanding drum to the floor. When you have reached your original location, gently raise the sanding drum and stop. You will now have made two identical passes, one forward, one backward, over the wood.

Move the sander four inches to your left and repeat the process . You will see that you are overlapping two or three inches. Overlapping two or three inches of the old surface will help you to gauge the depth of the cut. Continue this procedure until the entire width of the floor has been sanded.

After you have reached the end of the room lift the sanding drum; stop the machine, turn it 180 degrees and finish the other 1/3 of the floor. Over lap slightly where the 2/3 and 1/3 sections meet.

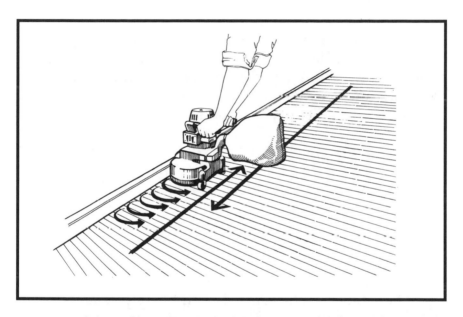

Use the power edger or random orbiting sander to sand up to the walls and other areas, like closets, that you could not sand with the drum sander. Where necessary use a sanding block and scraper. Be careful not to gouge the wood.

When using the edger, start at the wall, then work with the grain to the end of the drum sanded area. Then, pick up the edger go to the left and repeat the process. Overlap the drum sanded area slightly. You will still be able to see swirl marks where the drum sander and edge sander meet.

This first rough grit sanding should have removed the pre existing finish from an old floor or removed any marks and stains from a newly laid floor. Repeating the procedure with the medium and fine grit sandpaper is designed to eliminate the sanding scratches and provide the ultra smooth surface necessary for a beautiful wood finish.

In floor refinishing parlance, the entire first sanding with the rough sand paper is called the first cut.

Floor Repair

There may be some areas that have so much damage that the wood filler will have to be trowelled on the entire surface. If this is true with the floor you are refinishing, make the repair after the first cut. Trowel on the wood filler, clear away excess, and let dry thoroughly.

Some professional refinishers choose to make their own wood filler by combining saw dust from the first sanding with enough wood sealer to make a thick paste. The benefit to making your own wood filler is that you get an exact color match.

Second Cut

After all repairs have dried completely you are ready to make your second cut. Follow the same procedure for the second cut that you did for the first cut. Remember, the finish was removed with the rough sandpaper. You are just smoothing out the surface with the second cut. Typically, the second cut is with medium sandpaper, and the third cut is with fine grit sandpaper. The purpose of the fine grit sandpaper is to remove the scratches from the medium grit sandpaper. If the wood was in very bad shape, a fourth cut may be necessary.

Uneven Floors

When the floor is very uneven, you will have to make as many cuts as necessary to straighten out the floor. If the floor is very scarred, OFI recommends that the first and perhaps even the second cuts be made with medium, not rough grit sandpaper. Uneven, stressed wood is very prone to tearing and gouging. Use of the medium grit sandpaper keeps you in control and is easier on the wood. They also suggest that you make the cuts on a 45° angle to the direction of the flooring.

If the wood is so uneven that traditional sanding would remove a prohibitive amount of wood, you may have to use a 16" buffer with a very thick pad and sanding screen. The thick pad will conform to the wood's peaks and valleys.

Parquet, Block, Herringbone and Other Pattern Floors

Special care must be taken with patterned wood flooring. Never use coarse grit sandpaper on a pattern floor. Since the wood has been placed in a pattern, there is no way to go with the grain. The best way to sand these floors is to make the first two cuts on opposing diagonal lines.

Make your first diagonal cut with medium sandpaper. Then switch to fine sandpaper and make your second cut from the opposite diagonal. This is all the sanding that you should have to do with the drum sander unless the floor is in very bad condition. If additional passes are necessary, continue with the fine paper.

Final Cut

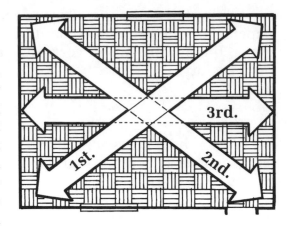

To finish the job take out all swirl marks with a buffer and fine sanding screen. This final cut should in a straight line across the length of the floor. If additional polishing is needed attach a non metallic steel wool pad to the buffer brush. Be careful not to over buff. If the surface becomes too smooth the stain will not be able to penetrate the wood.

Last Chance for Repairs

After the final sanding, inspect the floor carefully. This is the time to make any minor repairs that may have been exposed by the sanding process. If nails or nail holes are showing, now is the last chance you have to fill them. Be sure to counter sink each nail carefully. Use a matching color, commercial wood filler, such as Fix and Patch by the Darworth Co. or Smoothie Latex Wood Filler by Basic Coatings. Apply the wood filler with a putty knife; clear away excess; let dry completely, then hand sand.

The sanding job has not been completed until each room has been thoroughly vacuumed. Floors, walls, sills, every possible overhang. Then sweep down the floors with brooms covered by slightly water dampened cotton clothes.

Important!

By this time you have done a tremendous amount of work on these floors. Do not let anyone spoil them. It is best if you do not allow anyone to walk on a prepared floor except the finisher.

Never leave a sanded floor unprotected. It is as vulnerable as a newborn baby. Apply the first coat of stain, penetrating sealer, or urethane as soon as the area has been cleaned. You must begin the finishing process the same day that you complete the sanding.

23. 16" Buffer Technique

> **Materials Needed:** Open face coarse, regular medium and fine sand paper disks, fine sanding screen, nonmetallic steel wool pad for 16" buffer, coarse, medium and fine sandpaper disks for random orbiting sander.
>
> **Equipment Needed:** Respirator, 16" buffer, edger or random orbiting sander, sanding block, scraper, wet/dry, bag vacuum.

If the floor requires a great deal of leveling, start with coarse open grit sandpaper and follow the same steps recommended for drum sanding.

Sanding dust is going to be a very big problem. Fine oil based dust can be explosive, so make certain that the furnace, hot water tank, and all other pilot lights are off and no one is smoking while you are sanding. Tape shut the forced air vents and returns.

You can't sand what you can't see. If possible do this as a duo, one person sands, the other vacuums. Both of you should be wearing respirators and goggles. Be very careful not to hit the electric cord with the buffer. It is very easy to do.

It is a good idea to practice your technique before doing the actual work. Start without sand paper, in a kitchen or basement. When you have perfected your technique, put on the sand paper and start the actual job.

Because a buffer is so much lighter than a drum sander, you may have to repeat each step two or more times. When using the coarse and medium sandpaper, move very slowly and sand against the grain making 3' by 3' circles. Start in a corner at the wall and work going backward so that sanding dust will not obscure your vision.

Be very careful to keep a constant rhythm. Do not try to push the buffer in the direction you want it to go. A light pressure with the right wrist will send the buffer to the right. Pressure on the left wrist will send the buffer zooming over to the left.

Not paying attention, or pausing while the buffer is going can make the surface uneven. Buffers have a mind of their own and will try to run away from you.

You are using muscles that you never knew you had. Don't overtax yourself or you will ruin the job. Take regular rest breaks. Keep focused. Do not let yourself be distracted. Never rush.

When you get to the fine sanding cut, sand with the grain in a side to side motion moving the buffer very slowly. Go over each board several times to make certain that all the scratches put in by the medium paper have been removed.

If the floor is in very good condition, you may be able to eliminate the use of coarse sandpaper and use only medium and fine grit sandpaper.

If the floor has only had one light coat of finish, you may be able to get down to the bare wood with just one cut using a light sanding screen.

If you question how much finish is on the wood, make a test pass using just a light sanding screen. If that goes down to the bare wood, you have just saved yourself a great deal of work.

Remember, the art form when sanding is to take all the finish off the wood, but never to take off more wood than is absolutely necessary.

The sanding job has not been completed until each room has been thoroughly vacuumed. Floors, walls, sills, every possible overhang. Then "damp tack" the floors by sweeping them down with water dampened rags attached to a broom.

24. The Flecto Technique

Materials Needed: 3M 20 grit (very coarse), 36 grit (coarse), 60 grit (medium), and 80 grit (fine) sandpaper, 3M Scotch-Brite sanding pad, coarse, medium and fine sandpaper disks for random orbiting sander, wood filler.

Equipment Needed: Respirator, SQUAR BUFF orbiting floor sander, edger or random orbiting sander, putty knife, sanding block, scraper, wet/dry, bag vacuum.

The Flecto System is the new toy in town. The SQUAR BUFF sander is a 130 pound orbiting sander. This does not make it unique. KT's Planfix Finisher, to mention another, is a 95 pound, rectangular, orbital finisher. The SQUAR BUFF gives two handed control, wheels, double the sanding surface of the KT Planfix and automatic raising and lowering of the sander arm. In other words, it is heads and shoulders above the competition.

What makes this tool unique, is that where the other orbiting sanders are tools meant to supplement the drum sander, the SQUAR BUFF is meant to supplant it. The Flecto Verathane® people have discovered a market niche, "you, the consumer," and are trying to work it for all its worth.

Although the SQUAR BUFF is slower than the drum sander, in many ways it is the answer to a Do-It-Yourselfer's prayer. It is just about goof proof. It is light enough for a woman to use so a single gal is not at the mercy of us macho types. It can be used on any wooden surface.

The Flecto Varathane® people have produced a beautifully written and illustrated How To brochure entitled "Floor Finishing System," and designed a product display that houses all the 3M and Varathane® products necessary for the process. The brochure even tells you how much product you will need for different size rooms. All in all it makes floor refinishing pretty close to a "no brainer" which is just the way I like my DIY jobs to be.

PROCEDURE

Loading the SQUAR BUFF sander is very simple. Attach the 3M Scotch Brite pad to the Sander, then lower the sander on to one of the self adhesive pads.

1. Make certain that the floor has been thoroughly cleaned with a 4 oz. dry measure of TSP per gallon of water solution and rinsed several times. If the surface was waxed, a stripper should also be used.

2. For safety's sake, turn off all pilot lights and allow no smoking during sanding. Tape shut the forced air vents and returns. Always wear a respirator when sanding a large surface.

Sanding and vacuuming are equally important to successfully refinishing. The job will go a great deal faster if you can finagle a buddy to work with you. One sands, one vacuums. Both of you should wear respirators. Be sure to use a bag mounted wet/dry vacuum because the fine sanding dust can ruin the motor of a regular wet/dry vacuum.

3. The SQUAR BUFF people recommend that you begin the refinishing sanding procedure by scoring the surface with a medium sanding pad, then switching to the very coarse (20 grit) or coarse (36 grit) sandpaper to remove the finish.

The coarse paper may get clogged with finish residue. When that happens sanding efficiency goes way down and you will be producing a lot less sanding dust. Change to a new sheet of sandpaper immediately.

This is slow going and you'll find conversation difficult through a respirator and over the vacuum and sander. Be prepared to spend a lot of quality time with yourself. Think great thoughts about baseball, fishing, redecorating or cousin Alma's scandalous romance. You can't rush this. You can make the job less boring by trading off with the person doing the vacuuming.

4. When the finish is removed, smooth out the surface by resanding with medium (60 grit), then fine (80 grit) sandpaper.

5. After the final sanding and vacuuming, inspect the floor thoroughly. This is the time to make any minor repairs that may have been exposed by the sanding process. If nails or nail holes are showing, now is the last chance you have to fill them. Be sure to counter sink each nail carefully. Use a matching color, commercial wood filler, such as Fix and Patch by the Darworth Co. or Smoothie Latex Wood Filler by Basic Coatings. Apply the wood filler with a putty knife; clear away excess; let dry completely, then hand sand.

The vacuuming job is not completed until floors, walls, sills, every possible overhang has been cleaned.

6. Finally, sweep down the floors with a broom covered by a slightly water dampened cotton cloth.

25. Refinishing a painted hardwood floor to the natural look

Materials Needed: Citristrip™, garbage bags.

Equipment Needed: Old paint brush, 6" wide plastic scraper.

It is best to start a project like this knowing that you are in for a lot of work. Do not even think about sanding off a painted surface. The heat and abrasion of the sanding will make the paint gummy and clog sandpaper or sanding screens. It is best to remove the paint with a paint stripper first.

I recommend that you use a product called Citristrip™ by Specialty Environmental Technologies for this task. The fumes released by an ordinary stripper could be very dangerous. Citristrip is an environmentally friendly NMP based stripper. The house will smell like it is an orange juice plant. But the important thing is that everyone in the house will be able to breathe. My doctor tells me that breathing is very necessary for good all round health.

Citristrip is also very thick. This is very important because you will probably have to go through several coats of very tough floor paint. You can leave this product on up to 48 hours before it dries.

Your best bet is to spread the Citristrip on very thick with a paint brush. Put it on like your were frosting a cake (can you imagine a 9' X 12' pan cake?). Do not brush back and forth. Pat it on from one direction. It is best if you can put in on about 1/4" thick. You are going to be using a lot of paint stripper.

Let it sit over night. The next morning, scrape a small test patch. If the paint has lifted you can start removing the paint remover. If it has not lifted, let it continue working another 4, 6, 8, 12 hours. You are definitely waiting on the paint's time schedule, not your own.

Once the paint has lifted scrape it off with a wide plastic scraper attached to an extension handle. Make the scraper as wide as possible. A 6" blade would be wonderful. Scrape the gunk into plastic garbage bags and take it out to the garbage immediately.

If all the paint has not lifted, you may have to repeat the process.

Wash down the floor with mineral spirits.

Make any repairs to the floor that may be necessary.

Depending on the condition of the floor you may be able to just sand the surface with a 16" buffer and a light sanding screen. If more sanding is required follow the sanding directions above, for either the drum sander or the 16" buffer. Begin with the Second or Medium Sandpaper cut. If the floor is uneven, follow the directions for evening out the floor.

26. Can I refinish a factory finished hardwood floor?

Yes, but.... Refinishing some factory finished floors is a lot harder than refinishing a site finished floor. A site finished floor is wearing the equivalent of a cotton T shirt. An impregnated factory finished floor is wearing armor plating. Cotton is a lot easier to cut through than armor.

Prefinished floors may contain wax, silicone and ure-thane/wax combinations that have been impregnated into the pores of the wood. They may have had vinyl coatings added at the factory. You can not just "add" new coatings to these factory finishes.

Usually you don't have to. The majority of the manu-facturers of these extra tough finishes also make a renovator or reconditioner that can be used to restore the floor to its original appearance without the need for sanding.

Before deciding that refinishing is necessary give the floor a deep cleaning using a professional reconditioning products like Hartco Touch Up Kit in combination with To-tal Care or Pattern-Plus Shine, or Duraseal Renovator™.

27. Telling a prefininished from a site finished hardwood floor.

The boards in older prefinished floors usually have a distinctive 1/4" bevel along the outer edges. This bevel forms a shallow "V" groove between adjoining pieces of wood. Much of the newer prefinished flooring and most flooring that has been finished on the job site butt up together with no groove.

When you find that the floor has prefinished flooring, it is important that you sand, and finish both sides of the "V", in each groove, on every piece of wood. Not doing so, will at the least leave a splotchy looking floor. Many professional refinishers will not do grooves.

If you have any doubt about the wood being prefinished or your preparation of the wood, it is a good idea to do the refinishing job in miniature, on a closet floor, or other hid-den area. If the job goes wrong, you will be far happier redo-ing a small closet floor than having to strip and refinish a badly done living room, dining room, and hallway.

Refinishing Procedures

28. How do I refinish a hardwood floor?

We are redoing the hardwood floor in our baby's room and want to completely refinish. We like the natural color of the wood. What do we do?

Materials Needed: Easy Mask Painter's tape, Latex wood filler, tack rag, wood stain, Liquid Gloves™, 400 grit sandpaper, Fabulon Professional Crystal II, 100% cotton rags.

Equipment Needed: Push broom, long handled lambs wool pad applicator, paint tray, pole sander.

I'd suggest that even though you like the natural look you add a very light stain to the surface. The high sheen of the finish will change the "feel". Applying a very light stain will help provide a loving, warm feel to the floors that you want in baby's room. Trust me. Even my wife Barbara says that I'm right on this one.

Here's how to do the job:

1. Be sure you removed the quarter round molding, protected the wall molding with Easy Mask Painter's Tape. Clean and prep the room completely, per prior instructions.

If the floor was waxed, be sure to strip with a professional stripper. Wash away all grease and dirt with a solution of 2 oz. dry measure of TSP per gallon of water. Rinse several times with fresh water. Then dry thoroughly.

2. Completely remove the finish with a drum sander or a 16" buffer according to the directions given in tips # 22 and 23. Start with open face, coarse grit sandpaper for the major finish removal, then remove the sanding scratches with medium and fine grit sandpaper. Finish with the 16" buffer using a fine sanding screen and non metallic steel wool. Vacuum constantly as needed.

3. After you sanded the floor, vacuum thoroughly. This step has to be very thorough. Window sills, doors, ledges of any kind will hold dust. They all have to be vacuumed meticulously with the vacuum cleaner attachment.

4. When done, wipe down the floor with a slightly water dampened rag tied around a push broom.

   Let dry.

   It is vital that there not be even a speck of dust left after you have completed this step.

5. Now you are ready to stain. You can use Barkley, Minwax, Tripp or any one of a number of fine wood stains on the market.

   Wear refinisher's gloves or wear liquid gloves when you stain the floor. Use only a very light stain to accent the grain patterns in the wood.

6. For a top coat I would use Fabulon Professional Crystal II or other VOC water base floor finish. Many of the VOC finishes need a fortifier or additive added to the finish to make them adhere properly. Be sure that you read the can and make certain that you have all the components before you leave the store.

Apply one coat with a long handled lambs wool pad applicator, not a roller, or brush. The step seems deceptively simple. Many professionals say that DIY'ers tend to go too fast and do not apply even finish coats. Some suggest that a DIY'er might be better off applying the finish with a pad that will force them to go more slowly.

7. Let the first coat dry

8. Lightly sand the floor with 400 grit sandpaper or a fine sanding screen because the first coat will have raised the grain a little bit. Use a pole sander, so you do not have to bend over on your knees. Just lightly sand until it starts to powder.

9. Vacuum, then wipe the floor with a slightly water moistened 100% cotton rag.

10. Apply two or three more coats one hour to an hour and a half apart. Be sure to add the fortifier if you are using Fabulon or the coats will not adhere. You do not have to sand between these coats. Lightly screen, vacuum, and water tack between coats.

There are no harmful fumes with a VOC, and you can clean up with soap and water. It is very safe.

29. Staining Procedure

Staining is necessary with the new VOC urethane finishes because they are completely colorless and the stain "warms" the look of the wood at the same time that it accentuates the grain.

Apply the stain next to the wall molding with a brush and just tip it in to the edge. Be careful not to touch the molding. Wipe the stain on with a long handled lamb's wool applicator. Do not allow the stain to puddle. Spread in the direction of the grain. Wipe off the excess color with 100% cotton rags.

30. When should a hardwood floor be screened rather than sanded? Why?

It's not always an either/or situation. Sometimes you use both. There are basically four times when you use a sanding screen on a hardwood floor.

1. After you have completely sanded a hardwood floor down to the bare wood, there are usually marks where the drum sander got away from you, or swirl marks from the edging sander. A light screening removes these blemishes and provides a smooth, uniform surface.

2. On an older finished hardwood floor that does not need to be refinished, merely brightened with the addition of one or two new coats of finish. In this case you lightly sand with the sanding screen, to merely cut or "powder" the surface. This deglosses the old surface, creating a profile that will enable the new finish to adhere to the old.

3. The final time you screen a hardwood floor is after the first coat of finish has been applied. The first coat often raises the grain of the wood. This is especially true with the new VOC water base finishes. A light screening levels the grain and enables you to create a smooth finish with the second and third coats.

4. In hallways and rooms the flow of traffic has worn down a discernible pathway a sanding screen is used to "blend in" the area and prepare it for refinishing. Finish coats can not be added to areas that have not been screened.

The primary work is done on the well traveled path, then the rest of the surface is slightly screened. In this way you can add two or three coats to the worn area, then give the entire surface a couple of coats to even out the look of the finish.

31. Smoothing out a traffic pattern.

The front rooms of the house and hallways are all hardwood. How do I refinish the heavy traffic areas? How smooth do the floors need to be sanded before we can put on a new coating in the traffic areas? Also, can we change the overall look from a satin polyurethane to a high gloss?

Materials Needed: Equipment for removing quarter round molding, 3" Easy Mask™ painters tape, 100 grit sanding sheets for Flecto System SQUAR BUFF vibrating sander or fine sanding screen for 16" buffer, rags, stain if needed, Flecto Verathane Diamond or other crystal clear VOC acrylic urethane finish.

Equipment Needed: Equipment needed to remove quarter round molding, sanding respirator, Flecto System SQUAR BUFF vibrating sander or 16" buffer, random orbiting finishing sander, wet/dry vacuum cleaner, long handled lambs wool pad applicator.

This used to be a tough job. But now, thanks to a new vibrating machine that is part of the Flecto System, the job is definitely fast and easy. The only hard part is finding the machine. It is at many of the big home improvement centers. Flecto utilizes a vibrating system rather than a belt sanding system. The Flecto vibrating unit has a rectangular pad underneath and uses 60 to 80 grit sanding sheets to cut the top surface of the varnish.

If you have actually worn away the stain, so that there is a noticeably different color in the traffic pattern, you will have to restain that area. This is really a soul searching time. The stain you choose has to match the original stain completely, or the refinished area will not match the old. My advice is that you stain the area a little lighter. Then wipe some water on it for a glossy look. Dry thoroughly and decide if the match looks right. If not, add more stain.

Use one of the water base stains, such as Tripp wood stain, for this because they have a much faster drying time than oil base stains. If you use a water base stain you can go on to the next step in about four hours. If you use an oil base stain, you will have to wait a minimum of 24 hours.

1. Remove the quarter round molding from the edges of the floor. Use painters tape, not masking or duct tape, to protect the baseboard from being cut accidentally scarred by the vibration of the sander. My favorite tape for this job is Easy Mask™ by Daubert Chemical, because it is 3" wide and gives excellent protection, yet leaves no mark when it is removed.

2. Turn off the furnace, water tank and all pilot lights. Tape shut forced air registers and returns.

3. Scratch the surface with the vibrator just enough to make a profile. This just takes a couple of times back and forth. When you start to see it powder up, stop. At this point you are not disturbing the stain, you have only disturbed the top coat of floor finish. Use a hand held electric random orbit sander for the edges and corners.

4. After you sanded the floor, vacuum from top to bottom. This step has to be very thorough. Window sills, doors, ledges of any kind will hold dust. They all have to be cleaned meticulously with the vacuum cleaner attachment.

This is one of the most vital steps in the entire process. There should not be even a speck of sawdust left after you have completed this step.

5. Take a rag or towel slightly dampened with water and wash the floor. If this is a big job, you can attach the moistened rag to a push broom. Let dry.

6. Stain the bare spots in traffic pattern if needed.

7. Let dry 4 hours for latex stain, 24 hours for oil stain.

8. Apply two coats of a VOC clear acrylic finish to the floor. Use a lambs wool pad applicator to apply the finish. Do not use a roller. The lambs wool pad applicator is even better than a brush for this purpose.

9. Screen the floor lightly, vacuum and wipe down with a water dampened rag between coats.

If you choose the VOC Flecto or other VOC finish that you clean up with water, you can add the second coat the same day. The finish is clear as water, so you see the real color of the wood. Because it is a VOC, there are no fumes.

The VOC Flecto Elite Diamond Finish is a little bit different than what is presently on the hardwood. The present surface is probably an oil modified urethane. There will no longer be an "amberish" look. Add a minimum of two coats.

After you have added the new floor surface, keep it clean with frequent mop and vacuum cleaner cleanings. Specialized cleaners such as Poly Care™ by Absolute Coatings, Simple Green or Clear Magic. Mix the Simple Green and Clear Magic 10 oz. of cleaner to one gallon of water. Wash, and dry. Another product you can use is called "Wash Before You Paint". No rinsing is required.

32. How do I fix a dark pet urine, or any other, stain that has soaked into the wood and can not be removed with normal cleaners?

When I pulled up the carpeting from the floors of a house we just moved into, I found beautiful hard wood flooring. One room has a big, black, pet urine stain in the corner. It has really soaked into the wood. I've tried a TSP and bleach solution, everything. Nothing seems to work. What do I do?

Unfortunately, that room has to be sanded down to the bare wood and be completely refinished. Treat this room as a separate project that you do first. Then tackle the floors in the rest of the house.

Materials Needed: Alcohol based stain kill, like Bulls Eye B-I-N, wood bleach such as Klean Strip, Blanchett or Bondex, wood stain, Fabulon Professional Crystal II, 3M non metallic steel wool, light sanding screen.

Equipment Needed: Sanding respirator, drum sander or 16" buffer, random orbiting sander, goggles, refinishers gloves, sponge, paint brush, refinishers lambswool pad applicator and extension pole.

It is very important that you track down the urine problem completely or you will be haunted by the pungent ghost of cat every time the humidity rises. You know about the black urine stain on the floor. We are going to fix that with this tip. Just fixing the floor is not enough, thorough investigation is required.

Moisten the walls, especially low in the corners, and inspect them thoroughly both by eye and nose. If any odor is coming through the wall, the entire wall should be encapsulated with an alcohol based stain kill, like Bulls Eye B-I-N.

You will probably have some drips and overspray, so encapsulate the walls and edges before refinishing the floor. You will probably have to repaint the walls to cover the stain kill. Once the walls and edges are fixed let dry thoroughly, then use the following procedure to refinish the floor.

Use Bulls Eye B-I-N in the aerosol can to spray down into the edges where the flooring meets the wall and coat the under and wall side of the quarter round molding.

Unfortunately, you cannot just refinish a small portion of a floor with a dark urine stain. When the complete finish in an area has been removed, there is no way of matching the "patch" to the surrounding finish. Therefore the finish for the entire floor must be removed.

1. Clean and prep the room completely, per instructions at the beginning of Tip # 19, Page 32.

2. Strip the floor down to the raw wood. If you are lucky the sanding will take care of the stain.

3. Vacuum thoroughly. This step has to be very intensive. Window sills, doors, ledges of any kind will hold dust. They all have to be vacuumed meticulously with the vacuum cleaner attachment.

   After you have finished vacuuming, wipe down the floor with a slightly water moistened rag tied around a push broom. Let dry.

4. If the stain still persists, use a wood bleach such as Klean Strip, Blanchett or Bondex. Do not waste your time with laundry bleach. The best way to make the wood bleach very active is to heat the affected area with a heat lamp about 6" away from the wood for about 30 minutes. Remove the lamp and apply the wood bleach per instructions. If that does not do the job, repeat the entire process.

5. I hate to say this, but if you are not satisfied with the result, grit your teeth, and replace the offending boards. This is often the fastest way of solving the problem.

6. Finish the surface as you would a new floor.

33. I am removing the carpeting and going to leave the hardwood floors. How do I remove black rust marks from staples that were used to hold down the padding under the old carpeting?

Materials Needed: Citra-Solv™ or Foam Off by Biowash.

Equipment Needed: Mops and buckets, rubber gloves, goggles.

This fix will remove the marks but it uses a lot of water which is bad for the wood. Make sure that when you are finished cleaning off the staple marks you dry the floor thoroughly. Put a fan in the room to circulate the air and speed drying time.

1. Make a 4 to 1 mix of water and Citra-Solv™. Citra-Solv™ is a very heavy concentrate. There is no water in the bottle so you are making a very strong cleaning mix. Damp mop the floor with the mixture. Make a kind of slurry with your sponge mop. Keep it wet for at least a couple of minutes.

2. Rinse with a bucket of water and another sponge mop. Do the floor in sections. Do not rinse more than 1/3 of the floor before throwing away your rinse water and replacing it. This will do a very good job of cleaning the surface.

   If you use the Foam Off™, do not dilute.

1. Apply a liberal amount to the floor with a mop or roller. Let stand 15 or 20 minutes.

2. Remove with a wide bladed plastic putty knife.

3 Wash and dry the floor thoroughly.

34. How do I remove carpet padding foam or glue without destroying the finish?

Materials Needed: Old Hard Adhesive Remover™ or Foam Off™ by Bio-Wash Products.

Equipment Needed: Mop, rubber gloves, goggles, putty knife.

The probability is that you are going to damage the surface no matter what you do. Your best bet is to grit your teeth, work as carefully as possible, remove the foam or glue, then repair the finish if necessary.

Old Hard Adhesive Remover is very effective, but very harsh, has a terrible odor, and excellent ventilation is necessary. Make sure that you turn off the furnace, water heater, and all other pilot lights before using this product.

Foam Off was created for carpet foam removal, but will work its way through glue. It is relatively benign, has a citrus odor, and can be used with minimum ventilation.

1. Apply undiluted Foam Off to the foam padding. Let stand for 15 or 20 minutes.

2. Scrape off with a plastic putty knife. Be very gentle.

   If some of foam or glue remains, repeat the procedure.

3. When all the foam or glue has been removed wash with a clean mop and clean water. Dry with 100% cotton towels.

4. Inspect surface. If only a slight repair is necessary follow instructions for "Brightening" on page 17.

Floor Care & Maintenance

35. How do you care for a hardwood floor?

Hardwood floors and cats have a lot in common. They hate water and have to be groomed constantly.

> **Materials Needed:** Wash Before You Paint™.
>
> **Equipment Needed:** Elbow grease, vacuum cleaner, dust mop, sponge mop, and pail.

If you want to know how often you should clean your hardwood floor, just look at our friend the cat. A cat's favorite activity seems to be grooming itself. A hardwood floor wants the same amount of care, only you, not the floor, have to do it. A good vacuum cleaner is your best friend when it comes to daily maintenance. Vacuum the floor, then use a dust mop on every area you couldn't get to with the vacuum cleaner.

If somebody spills something, wipe it up immediately with a damp, not wet, rag.

Some hardwood flooring manufacturers go so far as to forbid the use of water on their floors.

If you feel you must use water every once in a while, damp mop. Never use an ammoniated cleaner. Not having to use soaps and various harsh chemicals is one of the benefits of your floor.

Dusting, vacuuming and an occasional very light clear water rinse is all you need. You are doing the right thing by keeping it clean.

The best product I've found for weekly cleanings is Wash Before You Paint™ by Culmac Industries. It comes in a little bottle of concentrate and a ready to use solution with a trigger sprayer. You want the concentrate.

A little bottle goes a long way. It is an all together different product. You can spritz it on and damp mop dirt off. It has a low pH so there is no residue and you don't have to rinse!

If it's Spring Cleaning time, or you want to make sure the floor is exceptionally clean for some reason there are many good special wood cleaning products on the market. Polywash by Absolute Coatings or Simple Green are both good maintenance products.

36. Heavy Protection.

We have three children and a 60 pound dog and just moved into a house with beautiful hardwood floors. What should we do to protect the floor?

Materials Needed: Minwax DuraSeal Cleaner and Sealer.

Equipment Needed: Dual 5" buffer.

Move. Seriously, children and big dogs are death on hardwood floors. You need a system that will require a little more maintenance but give a great deal more protection than most hardwood systems. I would recommend trying the DuraSeal system. This is a professional line by the Minwax Corporation. It consists of a DuraSeal cleaner and a DuraSeal sealer that you buff to a satin finish after it is applied.

I recommend a dual 5" wheel buffer, commonly called the Shetland buffer. It has twin 5 inch buffing pads and is relatively light weight and easy to use. This type of buffer is sold by Regina, Hoover, and many other companies.

37. Polyurethane finish sealers

I just bought an older home with wall to wall carpeting in the halls, family room, living room and dining room. When I pulled up the carpeting, I found that it was all beautiful hardwood flooring. It's in excellent condition. I want to keep the hardwood floors and just have an occasional throw rug. How do I seal and protect the floors to keep the hardwood looking beautiful?

Materials Needed: Rennovator™ by Minwax®, DuraSeal by Minwax, or the Hartco cleaning and sealing systems.

Equipment Needed: Sponge mops, dust mops, vacuum cleaner and attachments, dual 5" wheel buffer.

The first thing you want to do is deep clean the floor.

If you do not have heavy traffic, there are products like Renovator™ by Minwax® that you can use to deep clean the floor. Very good directions are right on the container.

For regular maintenance, vacuum and dust mop very thoroughly two or three times a week. Always vacuum first because the dust particles can scratch the floor. Use the dust mop to get into corners and under furniture that you can not reach with the vacuum cleaner tool.

Once a week you can damp (not wet) mop the floor. Make certain that you never permit a puddle of water to lay on the surface. It can make a mark that will cause a great deal of work to remove. Plain water in the bucket is fine. If your conscience won't let you believe the floor is clean without using a cleaner, add Wash Before You Paint, Polycare, or one of the other hardwood floor cleaners that are on the market.

Some people like to wax floors. I am not a wax proponent. However, professional flooring stores have a number of liquid and paste wax products. Minwax makes a pigmented wax, called Indian Sand, for use on floors which have areas that have been faded by the sun, etc. With paste wax, always remember: two thin coats are better than one heavy coat.

If you have young children, pets or a lifestyle that promotes heavier traffic, you should get into heavy duty hardwood cleaning and sealing systems like DuraSeal by Minwax and the Hartco cleaning and sealing system. It is important to use professional finishes because most consumer finishes are too soft and you have to work too darn hard to keep the floors looking good.

Both DuraSeal and Hartco product lines include a deep cleaner and a buffable sealer. These professional products give you a hard, protective surface that you can buff to a very good looking patina shine, but require minimal labor to maintain. No sanding is required and you can do the entire project in one day.

After using the deep cleaner, apply the sealer with a lambs wool pad applicator. Wipe up the excess sealer as soon as it is dry, buff the surface with one of those buffing machines with the twin 5" brushes. They are made by Hoover, Regale, Regina, and a number of other makers. You can rent these machine at many hardware stores. However, if you want to maintain your hardwood floors properly, it will be cheaper in the long run to buy one.

38. Should I wax or shouldn't I?

You have a right to be confused. There are some finishes that can not be waxed. If the finish on your floor is one of these, it will say so on the can.

Other than that, the world is divided into two camps. The Wax and the Don't Wax camps. I am a member of the Don't Wax camp.

As far as I am concerned, the only time a hardwood floor should be waxed is if the room is an almost unused guest room and the owner likes the hand rubbed, waxy look. Then, be my guest. No one will see it, but wax if you want to.

If you decide to join the Wax camp. On your head be it. You have to understand that once you start waxing, it is almost impossible to stop. It has to be waxed, buffed regularly, and stripped occasionally. In addition, the floor becomes more fragile. Water that is not wiped up at once will form water spots.

But why talk. If it is in your Karma that you are going to be a constantly working/worrying waxer, you are going to be a waxer and there is nothing I can do about it.

Go out and buy a good paste wax, a buffing machine, and buff your heart out. Enjoy!

39. Moving heavy objects over a hardwood floor.

How do I move heavy objects, like my refrigerator freezers or kitchen stoves without tearing up the finish?

Even if the refrigerator has wheels, it would probably be too much for the finish. The best way to move heavy appliances or any heavy object like television sets or pianos over hardwood is to buy 1/4" hardboard and make a hardboard path from the old location to the new location. Put the edge of the hardboard under the appliance and then roll the object on the pathway.

40. Chair protectors

How do I keep the chair and table legs from scratching the floor.

Applying adhesive felt protectors, made to fit the floor or table leg design will be a great help. Lift, never drag or push furniture over hardwood floors.

Problems

Bubbles or blisters

You goofed somewhere along the Finish/Refinishing process. Sometimes this is caused when the wood has not been cleaned and prepped properly. Usually the stain or one of the finish coats was not thoroughly dry before a new coating was applied. Sometimes this will happen if the floor was not screened between the first and second finish coat. Sadly, the only solution is to refinish.

Chewing gum

To remove chewing gum, wet it down with De-Solv-It™ by Orange Sol or Goo Gone™ by Magic American. Keep the gum moist with the product until it softens. Then wipe up with facial tissue. These products contain little or no water so they should not cause any white marks on the wood.

Cracking

Minute cracks appear in all wood. The cause is the opposite of cupping. There is too little moisture in the air. The best fix is proper humidification.

If you have hardwood floors, make certain that your furnace humidifier is at the proper setting. During the cold, high heating months, you should probably use a room humidifier to pump extra moisture into the air. Your floors, your house plants and sinuses will love you for it.

Cupping, wash boarding or peeling

The wood was flat and level when it was installed. However all wood will absorb moisture. If the edges are higher than the center of the board, commonly called cupping or wash boarding, the culprit is moisture. The fix is getting the humidity level of the entire house to its proper level. The problem took a long time to occur. Natural drying will take just as long.

It is important to realize that some cupping always occurs during the rainy seasons of the year. As soon as the humidity level gets down to normal the boards will flatten out.

In some severe cases you may have to completely resand and refinish the floor sanding off the raised edges. This sets you up for "crowning", when the board dries out. That is the center of the boards become higher than the edges. Your only solution is to completely resand the floor leveling the boards once again. This sets up a terrible sanding, resanding cycle that is not only hard on the person doing the sanding, but devastating to the floor. Your best bet is to let nature take its course and gradually straighten the boards.

Dull, blah wood floor.

For a quick facial for a dull but clean wooden floor, pour 12 oz. of Mop & Glow™ or Brite™ into a gallon of water and damp mop, then dry the floor. Either product will make the floor sparkle.

High heel shoe dents

I know some people who get so fanatic about their floors that they make everyone put on slippers. If you have the guts to tell your boss or your boss's wife to take off her high heels and put on slippers I want to watch. In the real world this is something you just have to live with.

Latex paint spills and spots

Use Goof Off II by Atlantic Sundries. Goof Off II was especially made for this job.

Roughness

If the surface is rough after refinishing, it means that the wood was just rough sanded. You can never just rough sand a wood surface. You must finish sand with medium, then fine grit paper. You must either live with this problem or refinish. If you decide to start singing the praises of the rustic or "colonial rough sawn" look, I will not tell a soul.

Scratches, early wear marks

This is usually a sign of improper care. Probable causes are tracking dirt into the house, not vacuuming with enough frequency, letting the pets nails grow too long.

If this gets too objectionable, you may have to rescreen the floor and add another couple of coats of finish. If family, friends and pets are too hard on the floor, you may have to upgrade to a super tough finish like Street Shoe.

Do everything possible to keep dirt from being tracked into the house. Make sure you are using entrance mats and area rugs. Increase the frequency with which you vacuum the floors. Be sure to put felt sliders on all chair bottoms.

Squeaky floors

If the floor was installed properly, the probable cause is improper humidification. Try increasing the humidity to 50%. If that doesn't work try graphite. The next step is to use products like Squeak Ender or shims tightening up the subfloor to the joists. If all else fails, use face nails, counter sink, and fill with colored wood putty.

Water stains

If a white water stain occurs, you may be able to clean it up with any brand of white tooth paste. Put a dab on a 100% cotton rag and try to rub out the stain. If that doesn't work, deep clean with Citra-Solv™. If the De-Solv-It™ alone does not do the job, shake a product called Rotten Stone, that you will find in most hardware stores, over the wet De-Solv-It™ and wipe the area. Use gentle pressure only. If the stain is still there you may have to re-screen and refinish the affected area.

Index

Selected Product Phone Numbers

Product Name	Company	Phone Number
Aqua Plastic	Coronado	800-883-4193
Blue Tape	3M	612-733-2895
Boen Hardwood		703-629-3381
Bondex Intl. Inc.		314-225-5000
Brite	S.C. Johnson	414-631-2000
Bruce Hardwood		800-722-4647
Bulls Eye B-I-N	Wm. Zinsser & Co.	908-469-8100
Citra-Solv	Bio-Wash (Canada)	800-663-9274
Citristrip	Specialty Envirnmental	810-340-0400
Clear Magic	Blue Coral Inc.	216-351-3000
Crystal II	Fabulon	716-873-2770
De-Solv-It	Orange-Sol Household	602-497-8822
Duraseal	Minwax Company Inc.	800-526-0495
Easy Mask	Daubert Coated Products	708-409-5125
Erickson's Flooring		810-543-WOOD
Fix and Patch	Darworth Co.	203-843-1200
Flecto System	Flecto	510-655-2470
Floor d-fenders	Equinox Products	414-473-2010
Foam Off	Bio-Wash (Canada)	800-663-9274
G & G Floor Co.		810-778-2050
Goof Off II	Guardsman Products Inc.	616-940-2900
Goo Gone	Magic American Corp.	216-464-2353
Gym Seal	McCloskey	800-345-4530
Hartco Touch Up Kit	Hartco	800-4 HARTCO
Hoover 5" Buffer	The Hoover Co.	216-499-9200
Indian Sand	Minwax	800-526-0495
Klean-Strip	W. M. Barr & Co. Inc.	901-775-0100
Non Metallic Steel Wool	3M	612-733-2895
Oak Flooring Inst.	NOFMA	901-526-5016

Old Hard Adhesive
  Remover .... Tile Helper Inc. 708-453-6900
Pacific Plus .... Bona Kemi USA Inc. 303-371-1411
Pacific Strong .... Bona Kemi USA Inc. 303-371-1411
Pattern Plus .... Hartco .... 800-4 HARTCO
Paynter Floors 810-471-9090
Planfix Finisher .... Kunzle & Tassen 713-728-3800
Polycrylic .... Minwax 800-526-0495
Quick-Dry
  Wood Stain .... Basic Coatings 800-247-5471
Renovator .... Minwax 201-391-0253
Respirator
  Dust/Mist
  Sanding 3M 800-247-3941
Safe & Simple .... Carver Tripp 800-225-8543
Sikkens TFF .... Akzo Coatings Inc. 800-833-7288
Simple Green .... Sunshine Makers Inc. 800-228-0709
Smoothie Latex
  Wood Filler .... Basic Coatings 800-247-5471
Squar Buff Sander Flecto Co. Inc. 510-655-2470
Squeak Ender........ E & E Engineering Inc. 810-978-3800
Street Shoe
  Commercial
  Wood Finish .... Basic Coatings 800-247-5471
Scotch-Brite
  Sanding Pad........ 3M DIY Div. 612-733-2895
Synthetic
  Steel Wool........... 3M DIY Div. 612-733-2895
Sanding Screen..... 3M 612-733-2895
Total Care .... Hartco Floor Care
                          Products .... 800-4 HARTCO
Varathane Elite
  Diamond .... The Flecto Co. Inc. 510-655-2470
VOC Flecto
  Varathane .... The Flecto Co. Inc. 510-655-2470
Wash Before
  You Paint .... Cul-Mac Industries Inc. 800-626-5089
Zip-Guard
  Wood Finish .... Star Bronze Company 800-321-9870

Ask for these titles at your favorite bookseller.

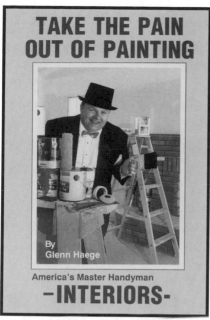

Take the Pain out of Painting - Interiors- by Glenn Haege

Take the Pain out of Painting - Exteriors- by Glenn Haege

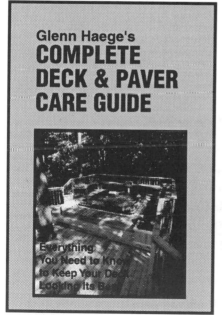

Glenn Haege's Complete Deck & Paver Care Guide

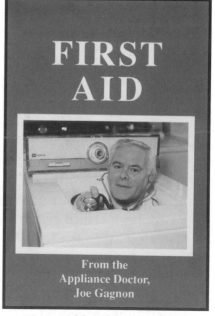

First Aid by the Appliance Doctor, by Joe Gagnon

Learn How To *Paint it* Inside:

Take the Pain out of Painting -Interiors- by Glenn Haege

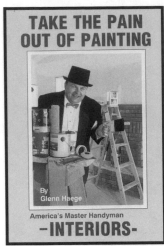

ISBN 1-880615-19-3
$14.97

Even people who have been painting for twenty years or more tell America's Master Handyman, Glenn Haege, that they started painting faster, easier, and with better, more professional results after reading this book.

The Bookwatch says:

"Finally: a guide to interior do-it-yourself painting which follows a very simple yet information packed step-by-step format!...A very basic, essential home reference."

The Detroit News says:

"Haege makes it easy for anyone smart enough to lift a paint can lid... to solve a particular painting problem."

This one, power packed book contains the information you need to have a great looking job every time.

This book will show you how to:
- **Prepare a room so well you may not need to paint.**
- **Remove wallpaper and peeling paint easily and prepare a firm painting foundation.**
- **Paint even slippery surfaces like kitchen cabinets.**
- **Use special Stain Kill Paints to solve *impossible* painting problems.**
- **Make paint look like wallpaper, wood, and stone, in a fabulous 40 page "Faux Finish" section.**
- **Plus Much, Much More.**

Fast & Easy! © *1995 MHP*
Outside:
Take the Pain out of Painting -Exteriors- by Glenn Haege

Your home's exterior paint is all that protects your family's largest single investment from the elements. America's master handyman, Glenn Haege, shows you the easy, economical ways to give your house the protection it deserves.

ISBN 1-880615-15-0
$12.95

George Hampton of *The Booklist* says:
"Writing for paint retailers, contractors, and homeowners alike, Haege vends plenty of practical information organized into step-by-step procedures for everything"

Hormer Formby, the originator of Formby Finishes says:
"Glenn Haege knows more about Paints and Products than anybody I have ever known."

This book will show you how to:
- Get the most for your painting dollar.
- Choose a painting contractor.
- Get rid of mold and mildew.
- Prepare the surface so the paint will wear like iron.
- Paint wood, vinyl, aluminum, concrete, log, shingles and all other exterior surfaces.
- Varnish or revarnish a log cabin.
- Paint, stain, or varnish exterior toys and furniture.
- Plus Much Much More.

Fix it Fast & Easy!
with a Little Help from America's Master Handyman, Glenn Haege

ISBN 1-880615-00-2
$14.95

DIY technology changes daily. Problems that would have been almost impossible to solve just a few years ago, can be fixed quickly and easily, today. Glenn Haege can show you how. He has been solving America's DIY problems on the radio for over 12 years. Now, the very best of these tips, are included in this collection of most asked "How To" questions.

Kathleen Kavaney Zuleger's Book Review column says:
"The book tells the easiest way to do many of the hardest cleaning and fix-up chores. Haege names names, and tells the reader which products will do the best job."

The Bookwatch **says:**
"From handling mildew problems to revitalizing a deck and sprucing up furnitue, this book has hints others miss....It is these tips on common yet seldom-addressed problems which make this such an important home reference."

There are special sections on Cleaning, Walls, Floors, Exterior Cleaning & Painting, Furniture Renewal & Repair, Decks and Heating.

This book includes:
- **Solutions to mildew problems.**
- **How to get professional painting painting results.**
- **Easy ways to remove water rings from furniture.**
- **Furniture refinishing tips.**
- **Cleaning dingy aluminum siding.**
- **Plus Much, Much More.**

Vital, Need-to-Know Info
for Deck Owners

Whether you are one of the 30 million Americans who own a deck, or are one of the one million who will build a wood deck or paver patio this year, this guide contains vital, new research you won't find anwhere else.

Glenn Haege's
COMPLETE
DECK & PAVER
CARE GUIDE

USDA Forest Service research indicates former staining and sealing recommendations may be wrong. Canadian proprietary research indicates some of the most popular deck cleaners may actually damage decks.

ISBN 1-880615-39-8
$4.95
Available June 1995

The TV commercials are very confusing. How are you supposed to know which products will do the best job on your deck or paver walk or patio? America's Master Handyman, Glenn Haege, has done the research and compiled easy to understand instructions for you.

Here's a list of some of the information in this guide:
- **How to decide whether a deck or paver patio is best for your lifestyle.**
- **The differences in deck woods that can save you thousands of dollars.**
- **How to keep pressure treated wood from splitting.**
- **How to cut through the confusion about sealers, toners, UV coatings and wood stains.**
- **How to create a "furniture finish" deck.**
- **Easy step by step instructions on deck & paver care.**
- **How to bring back the beauty to deck wood.**
- **How to make paver colors come alive.**
- **How to keep pavers looking beautiful and new.**

First Aid for your pocket book.

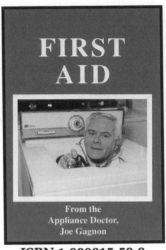

FIRST AID

From the
Appliance Doctor,
Joe Gagnon

ISBN 1-880615-50-9
$14.97

This book is for you if you would rather spend your money on a new cd player than replacing the garbage disposer.

You don't get any medals for repairing or replacing a major appliance. *Joe Gagnon* shows how to keep them running better, longer; how to repair them when they break, and how to cut through the hype when it comes time to buy new.

The Chicago Sun Times says:
"Many Home Owners will save money with a copy of *First Aid From the Appliance Doctor*, by Joseph Gagnon. They also could save some lives."

The Davis Enterprise says:
"To become empowered, read appliance doctor's book."

TWENTYONE Magazine says:
"The information is *First Aid* can save you money. It's a book every homeowner will want to keep on the reference shelf."

This book will show you how to:
- **Save hundreds of dollars on appliance purchase and repair.**
- **Cut through the lies in retail appliance ads.**
- **Tell the difference between a cheap promotional appliance and one that's built ot last.**
- **Make your appliances run better and last longer.**
- **Master the repairs you can easily do yourself.**
- **Keep from being ripped off on parts and service.**

TO: **Master Handyman Press, Inc.**
P.O. Box 1498
Royal Oak, MI. 48068-1498

Please send me copies of the following books:
All books are sold with a 100% money back, satisfaction guaranty:

___ **FIX IT FAST & EASY!**   @ **$14.95 each = $** _____
___ **TAKE THE PAIN OUT OF PAINTING!**
      - INTERIORS -   @ **$14.97 each = $** _____
___ **TAKE THE PAIN OUT OF PAINTING!**
      -EXTERIORS-   @ **$12.95 each = $** _____
___ **Glenn Haege's COMPLETE DECK &**
      PAVER CARE GUIDE   @ **$ 4.95 each = $** _____
___ **Glenn Haege's Complete HARDWOOD**
      FLOOR CARE GUIDE   @ **$ 6.95 each = $** _____
___ **FIRST AID from the Appliance Doctor,**
      Joe Gagnon   @ **$14.97 each = $** _____

           Total $ _____

Michigan Residents: Please add 6% Sales Tax.

Shipping: Surface $2.50 for the first book and $1 for each additional.
Air Mail: $3.50 per book.

       SHIPPING: _____

           Total $ _____

Name: _____

Phone No _____

Address: _____

_____ ZIP: _____

Credit Card Information. Please fill out if you wish to charge.

Please charge my _____ Visa _____ Master Card

Expiration Date: _____ Card # _____

Name on Card: _____

Signature: _____

Mail to:

Master Handyman Press, Inc.
P.O. Box 1498
Royal Oak, MI 48068-1498